No Sweat Multiplication

Knowing and Understanding Multiplication & Division Facts

by

Carol Pirog

Anchor Book Press · Palatine

No Sweat Multiplication: Knowing and Understanding Multiplication Facts

Anchor Book Press
440 W Colfax Street, Unit 1132, Palatine, IL 60078
ISBN: 9781949109

Printed in the United States

Introduction

I firmly believe that the biggest problem for students struggling with math is a lack of understanding of the prerequisites for the math they are currently plagued with. After all, if a student lacks a basic understanding of the base ten system, he or she is going to have difficulty reading numbers, regrouping, decimals, etc. If a student does not have a basic understanding of fractions, understanding decimals is really going to be difficult. The list could go on and on. Not only that, in talking with high school math teachers, one of the biggest complaints I hear is that students coming to them do not know multiplication and division – the facts or the concepts.

There is a big controversy over the issue of whether students actually need to memorize multiplication and division facts. While it is not necessary to put pressure on students to complete facts in record time as all students process information at differing speeds it is important that students know the facts. Multiplication and division are essential skills as they are the foundation for knowing the relationship between numbers. This is necessary for higher level math. Take fractions for example, it is just way too cumbersome for a student to add or subtract fractions with unlike denominator if they have no clue what the multiples are of the denominators. After all, how do you choose the lowest common multiple if you do not know the multiples. Further, research has shown that students who are the most proficient in math have memorized math facts and learned math concepts at the same time. Memorization is necessary for math fluency. If a student spends so much time figuring out facts, it distracts from the main problem and makes it more difficult to understand. On the other hand, if a student memorizes without understanding, the information is not helpful when needed in another situation or calculating complex problems.

For many students, division is harder than multiplication. I believe that is because multiplication is taught first. Therefore, multiplication is easier because students have more practice, not because it is an easier concept. If you are not sure, think of this analogy. Most children totally understand the concept of division in real life situations. They know if there are 10 pieces of candy and it has to be shared equally, they will be dividing the candy between two kids. Yet, when we put division problems on the paper, they struggle. Not only do they struggle with the answer, they struggle in creating an equation, in knowing which number to start with. Learning facts will help a child become just as confident with division as with multiplication.

And briefly, for those worried about students with memory issue, unless the student gets lost every day on the way to class, he or she will be able to memorize facts if not presented with an unreasonable number of facts at one time.

An investment in knowledge pays the best interest.

Benjamin Franklin

What Do the Standards Actually Say?

There has always been a lot of disagreement over whether students need to learn facts or not. With Common Core Math, one of the ideas was that students would understand what they were doing rather than just rote memorization of math facts. But for years, learning math facts was a standard that students were expected to learn. In addition, prior to Common Core, the curriculum activities were centered around teaching students the concept of multiplication and memorizing math facts. Having said that, unbeknownst to many teachers, parents, and students, Common Core does not say students no longer need to memorize the facts. Let's look at what the 3rd grade standards actually say.

3.OA.C Multiply and divide within 100.

- **3.OA.C.7 Fluently multiply and divide within 100, using strategies such as the relationship between multiplication and division (e.g., knowing that 8 × 5 = 40, one knows 40 ÷ 5 = 8) or properties of operations. By the end of Grade 3, know from memory all products of two one-digit numbers.** (The Common Core standards are available on state education sites that have accepted Common Core as well an many other places – just Google it for more information.) If your state has not accepted Common Core, they will have a similar standard on the state education site as all states teach multiplication and division in math. It may be in the 3rd grade standards or the 4th grade standards.

When standards say **fluently**, this indicates students should know facts without counting on fingers or using other methods to calculate the answer. Students know the answer because they have memorized it. Why is this needed? Most of us are familiar with reading fluency – the ability to quickly read a passage, sounding out only 1 or 2 unknown words. Fluency in reading allows a student to use available brain resources for understanding the meaning of the passage (reading comprehension) instead of decoding words. This does not mean the student has not learned strategies to decode unknown words. The use of **fluency** in this 3rd grade Common Core Math Standard shows us that **fluency** in math, while not as well known, is also expected. Math **fluency** plays the same role in math as reading fluency plays in reading. **Fluency**, in math and reading, improve comprehension. Students working difficult math problems are distracted from the problem, if they have to take time to figure out facts. While students who are **fluent** in math facts can focus on problem solving.

What then is the difference between the rote memorization of facts 50 years ago and the memorization of facts today? Nothing, except Common Core memorization is not called 'rote memorization.' Rote is a descriptor added to give memorization a negative connotation. While 'rote memorization' has a specific meaning, it is used indiscriminately by many people without regard to details that make a difference. Therefore, the goal of Common Core math is largely misunderstood. It is not to get rid of memorization, but to get rid of **rote** memorization which is memorization without understanding. As far as facts go, the facts were never meant to be taught without attempting to teach understanding. Look through old math books and you will see the pages with pictures clearly meant to help students understand the concept of multiplication. Common Core wants students to memorize facts and understand multiplication, as was the goal fifty years ago. How do we know? It is a simple matter to check the Common Core Standards. The Standard on the previous page says kids need to memorize facts. While the Standard which is below is about understanding multiplication. The problem comes when teachers and administrators lump all memorization under the category of rote memorization and teachers in other grades are not aware the standard calls for memorization.

3.OA.A.1 Interpret products of whole numbers, e.g., interpret 5 × 7 as the total number of objects in 5 groups of 7 objects each. (Quoted from Illinois Board of Education Website)

- Count equal groups (3-E.1)
- Identify multiplication expressions for equal groups (3-E.2)
- Write multiplication sentences for equal groups (3-E.3)
- Relate addition and multiplication for equal groups (3-E.4)
- Identify multiplication expressions for arrays (3-E.5)
- Write multiplication sentences for arrays (3-E.6)
- Make arrays to model multiplication (3-E.7)
- Write multiplication sentences for number lines (3-E.8)
- Relate addition and multiplication (3-N.9)

When we look at this standard, the first standard for multiplication for 3rd graders is about students understanding multiplication. So, rather than eliminate memorization of facts, Common Core includes understanding of multiplication with the memorization of facts. People who know math, know that multiplication and division facts are the basis for understanding the relationship between numbers. Because of that, fluency (memorization) of math facts is essential to the understanding of higher-level math concepts.

The purpose of this book is to help your child become fluent in multiplication facts as they learn the concept of multiplication and division. This will build a solid foundation so he or she can become competent in understanding and solving higher-level math.

To solve math problems, you need to know the basic mathematics before you can start applying it.

Catherine Asaro

How to Use this Book

Key Concepts for Parents (Teachers can find classroom use at the end of the book):

- Fun, fun, fun – do all you can to keep this fun – rewards, competition, charts
- A child needs to pass one level before moving to the next level – high five, sticker, smiley face – an acknowledge each accomplishment
- Teach new facts the day before doing the student practice page; figuring out answer using understanding of multiplication concepts before practicing fluency. New Facts Chart in back shows new facts for each lesson
- Student should know facts from memory when they do the student practice page (Lessons A – Z and Mastery 1 – 5)
- If student is on any level for more than 2 or 3 days, practice new facts in isolation (Note: it is more effective to practice for 2 minutes 5 times a day than it is to do 1 ten-minute practice session.)
- Students can complete in writing or orally.
- Teach vocabulary – if you need more help, check out *Teaching Vocabulary for Common Core Operations: Addition, Subtraction, Multiplication and Division*
- One-size does not fit all, use additional strategies if you child or students still struggle. (See back of book for additional strategies.)
- If using paper and pencil, copy page for child to use.

The top section of student practice page (Lessons A – Z) is for understanding. The bottom section is memorization and should be timed. Depending on age and manual dexterity of your child, anywhere from 20 – 30 seconds should be enough. Timing is to keep student focused, not to measure speed. Students can answer orally. Although, research has shown that the physical act of writing triggers the memory area of the brain. Unless your child has a physical handicap, which makes writing difficult, writing will make the memorization process faster.

Key Ideas (teach these concepts) **for Student:**

- Multiplication facts that are reversed have the same answer – 3 x 5 has the same answer as 5 x 3
- 3 x 5 is read as 3 times 5
- 3 x 5 means 3 groups of 5 objects
- Multiplication is repeated addition so 3 x 5 equals 5 objects + 5 objects + 5 objects = 3 groups with 5 objects in each group. 3 x 5 = 5 + 5 + 5

A person who never made a mistake never tried anything new.

Albert Einstein

Let's Get Started with Multiplication

Before starting student practice pages work with child on understanding multiplication:

Day 1 (Understanding multiplication is finding the total number in a set of equal groups)

1. Work with manipulatives to help child understand – use counters, pennies, M & Ms, etc.
2. Start with 3 x 5. Put 5 objects in each of 3 groups (make groups by using bowls, saucers, or circles drawn on a sheet of paper, etc.). Have child count all objects (if your child cannot count to 15, he or she is not ready for multiplication. Write on paper or white board, "3 groups with 5 objects in each group equals 15". Say, "The math equation to go with this sentence is 3 x 5 = 15." (Be sure your child knows the meaning of the word multiplication and the meaning of the word equation.)
3. If we make five groups instead of 3, how many will be in each group (if necessary, prompt by saying the same number must be in each group). That's right, 3. If we put our 15 objects into 5 groups, each group will have 3 objects. So, 5 x 3 = 15. Draw a picture of both facts (15 objects in 3 groups and 15 objects in 5 groups. Explain to your child that it is the same number of objects, just grouped differently).
4. Play around with some other problems – **3 x 4** or **2 x 6** or **3 x 6** (keep the numbers small, so your child is not struggling with other concepts and can focus on multiplication). When teaching understanding, it is best to use small, easily manipulated facts. This is so your child understands what is happening when he or she multiplies. Understanding is the difference between rote memorization and Common Core goal of fluency.
5. Introduce **fact families.** Work with 3 x 5. Tell child that there are families of facts. A fact family is a group of three numbers that can make 2 multiplication facts and 2 division facts. So, 3, 5, and 15 are a family. Draw your child's attention to the fact that the multiplication problems end with the largest number because multiplication finds the total. The division problems start with the largest number because division is separating the total into equal groups.

   3 x 5 = 15
   5 x 3 = 15
   15 ÷ 3 = 5
   15 ÷ 5 = 3

Note: If you child needs to spend more that one day on any of these lessons, it is better to spend the time now, than rush to get through the material. The goal is a solid foundation. For younger children it might be good to do points 1 and 2 the first day. Point 3 on the 2nd day. Point 4 on the 3rd day. Point 5 on the 4th day.

Day 2 (Multiplication is Repeated Addition)

1. Put the counters in 3 groups of 5. Ask your child if he or she can write an addition problem. If not, show him or her - 5 in this group + 5 in this group + 5 in this group or 5 + 5 + 5 = 15, which gives us the same answer as 3 x 5 = 15.
2. Repeat with 3 x 4. This time your child should be able to write the addition problem of 4 + 4 + 4.
3. Try with 2 x 5 and 2 x 6.
4. Review on Day 3 if your child seems to struggle with the concept.

Day 3 (Zero Property of Multiplication)

1. Work with numbers multiplied by 0.
2. 3 x 0 (Explain: You are going to have 3 groups with zero (nothing) in the groups. Ask you child how many objects. Some children struggle with this as they think there has to be some if you are asking. Simply explain, it is zero. We have no objects.
3. Try with 2 x 0.
4. Now tell your child there will be 5 objects in each group, but you don't have any groups. Write the problem for him or her as you say, 'zero groups with 5 objects in each (0 x 5). We have no objects because we have no groups. Sometimes this is easier to understand using concrete objects, such as gum, candy, or cars. For example: 5 Matchbox cars come in a pack. How many Matchbox cars do we have on the table (be sure you have removed all cars from the table). Right, we have zero (zero is the number that represents none) because we do not have any boxes of Matchbox cars on the table. The problem is 0 boxes times 5 cars in a box equals 0. Likewise, the equation for this problem is 0 x 5 = 0.

Day 4 (Identity Property of Multiplication)

1. Teach the identity factor in multiplication. Have your child say, **identity property and identity factor** (make sure your child knows what a factor is). Talk about what the word identity means. Your identity is (mom, dad, teacher, etc). Ask what his or her identity is (prompt with his or her name, if necessary). Then write the letter A on paper. The identity of that letter is a. Its name is A. Do a couple more letters and then switch to numbers. Once your child knows the identity of a number is simply what it is called, you can move to the next step.
2. Tell your child, in multiplication there is a number called the **identity factor** - when you multiply a number by the **identity factor**, you get that same number. For example, 1 x 3 = 3 (use the manipulatives). Do a couple more numbers then ask, what do you think 1 x 53 is? If your child is struggling with this idea, make 1 circle explaining you have 1 group. Then write the number 53 in the circle. Ask how many in the group? Some children will generalize quickly, others need to walk through the concept. Do as many 2-digit numbers

as you need in order for your child to understand. Then move to 3-digit numbers. Make a big deal of doing really big numbers, like 3 million, 24 billion, etc. This concept is known as the **Identity Property of Multiplication.**

Day? (Rule of 10)

(There is no day here because you should come back to this once you feel your child has a firm grasp of the concept of multiplication and knows quite a few facts including 4 x 10.)

1. Ask your child what the multiples of ten are (you are looking for 10, 20, 30, 40, 50). You might need to teach the definition of multiple. Use circles with the number 10 in each if you need a concrete example. Then count by 10s, pointing to each circle as you count.
2. Ask what is 1 x 10; 2 x 10; 3 x 10; 4 x 10. Write these problems going down on a sheet of paper, one under the other. Circle all the answers. Ask your child how the answers are alike (they all have zeros). There is a pattern. If your child suggests another pattern besides the fact that they all have zeros, just prompt them to see what other patterns they can see.
3. Look at each answer; and then talk about the problem. What do you think 8 x 10 will be? (Hint for your child: what number was in all the answers – right, zero). Do you think this answer will have a zero? What other number do you think it will have if 1 times 10 is 10 and 2 times 10 is 20. So, 8 times 10 will be? ___
4. What is 9 x 10? Right, 90. If your child needs help with 80 and 90, come back to this lesson next week.
5. What do you think 13 times 10 would be? (Hint: we know it is going to have a 13 and a zero, lets write it down). Good, what number do you have? Right, 130. How about 27 – write it if it is easier.
6. The pattern is to add a zero, we know that because the rule for multiplying by 10 is the answer ends with a zero. This lets us figure out really big numbers when we can't count by 10s.

Day? (Rule for 100)

1. Teach rule for 100. Follow the Rule for multiplying by ten, changing where necessary to fit the pattern for 100 times a number.

Level A

A

Draw a picture:

Example: 5 x 0 2 x 0

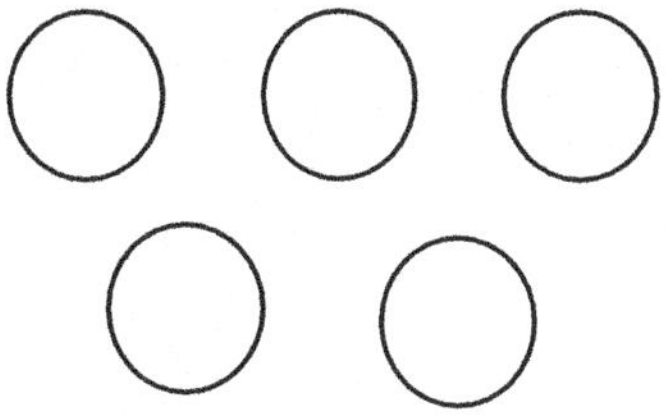

Note: There are no Xs because it is 5 groups of zero objects in each group

Write in a word sentence:

Example: 3 x 0 – <u>three groups with zero objects in each</u>

4 x 0 ________________________________

Zero Property of Multiplication: Any number times 0, equals 0 **A**

2 x 0 = _____	5 x 0 = _____	9 x 0 = _____	0 x 1 = _____
6 x 0 = _____	0 x 7 = _____	3 x 0 = _____	0 x 0 = _____
0 x 8 = _____	0 x 10 = _____	7 x 0 = _____	8 x 0 = _____
0 x 3 = _____	0 x 4 = _____	0 x 6 = _____	0 x 5 = _____
0 x 2 = _____	4 x 0 = _____	1 x 0 = _____	10 x 0 = _____

Level B

B

Draw a picture:

Example: 5 x 1

3 x 1

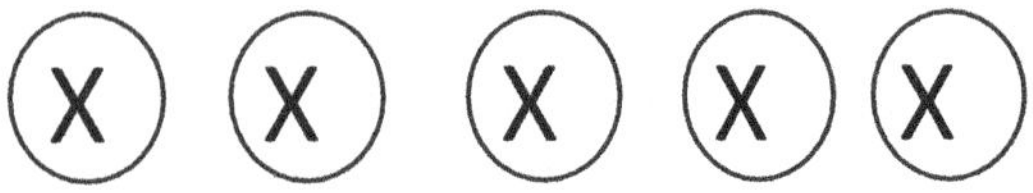

Note: There is one x in each group

Because it is five groups with 1 object in each group

Write in a word sentence:

Example: 4 x 1 – four groups with one object in each

7 x 1 ______________________________

Identity Property of Multiplication: if you multiply a number by 1, you will get the same number

B

2 x 1 = _____	0 x 1 = _____	9 x 1 = _____	1 x 1 = _____
6 x 1 = _____	1 x 7 = _____	3 x 0 = _____	0 x 1 = _____
1 x 8 = _____	1 x 10 = ____	7 x 1 = _____	8 x 1 = _____
1 x 3 = _____	1 x 4 = _____	1 x 6 = _____	1 x 5 = ______
0 x 2 = _____	4 x 0 = _____	1 x 1 = _____	10 x 1= _____

Level C

C

Draw a picture:

Example: 4 x 2 3 x 2

Note: There are 2 Xs in each group

Because it is four groups with 2 objects in each group

Write in a word sentence:

Example: 6 x 2 – six groups with two objects in each

10 x 2 ______________________________

New Facts: 2 x 2 2 x 4 4 x 2 **C**

7 x 1 = _____	0 x 8 = _____	1 x 3 = _____	2 x 4 = _____
2 x 2 = _____	1 x 5 = _____	4 x 2 = _____	1 x 1 = _____
1 x 4 = _____	3 x 0 = _____	0 x 1 = _____	1 x 8 = _____
2 x 4 = _____	1 x 9 = _____	4 x 2 = _____	2 x 1 = _____
10 x 2 = ____	2 x 2 = _____	6 x 1 = ______	0 x 6 = ______

Level D

D

Draw a picture:

4 x 2

3 x 2

Write in a word sentence:

Example: 3 x 2 – <u>three groups with two objects in each</u>

10 x 2 ______________________________

New Facts: 2 x 6 6 x 2 2 x 8 8 x 2

D

0 x 9 = _____	2 x 8 = _____	1 x 3 = _____	6 x 1 = _____
2 x 6 = ____	2 x 4 = _____	1 x 5 = _____	4 x 2 = _____
1 x 1 = _____	6 x 2 = _____	1 x 4 = _____	7 x 0 = _____
8 x 1 = _____	8 x 2 = _____	2 x 2 = _____	7 x 1 = _____
1 x 9 = _____	4 x 2 = _____	2 x 1 = _____	10 x 1= _____

Level E

E

Draw a picture:

2 x 6

5 x 1

Write in a word sentence:

1 x 8 ______________________________

3 x 2 ______________________________

New Facts: 2 x 3   3 x 2   2 x 10   10 x 2

E

2 x 4 = _____	2 x 3 = _____	4 x 2 = _____	2 x 10 = _____
1 x 7= _____	0 x 5 = _____	2 x 8 = _____	0 x 3 = _____
2 x 2 = _____	2 x 6 = _____	2 x 1 = _____	10 x 2 = _____
9 x 1 = _____	3 x 2 = _____	6 x 2 = _____	1 x 4 = _____
6 x 1 = _____	8 x 1 = _____	8 x 2 = _____	2 x 2 = _____

Level F

F

Draw a picture:

4 x 2

3 x 0

Write in a word sentence:

9 x 2 – ______________________________

1 x 7 ______________________________

New Facts: 2 x 5 5 x 2 2 x 9 9 x 2

F

2 x 5 = ______	2 x 8 = ______	9 x 2 = ______	2 x 2 = ______
2 x 6 = ______	2 x 4 = ______	2 x 3 = ______	4 x 1 = ______
2 x 10 = _____	2 x 9= ______	2 x 4 = ______	10 x 2 = _____
4 x 2 = ______	2 x 3 = ______	6 x 2 = ______	5 x 2 = ______
6 x 1 = ______	8 x 1 = ______	8 x 2 = ______	0 x 2 = _____

Level G

G

Draw a picture:

3 x 3

Draw an array:

4 x 2

Example: X X X X
X X X X

4 x 3

Write in a word sentence:

7 x 2 ________________________________

2 x 5 ________________________________

New Facts: 3 x 3 2 x 7 7 x 2

G

5 x 2 = ______	3 x 3 = ______	8 x 0 = ______	3 x 2 = ______
2 x 2 = ______	1 x 5 = ______	2 x 8 = ______	7 x 2 = ______
2 x 7 = ______	1 x 9 = ______	2 x 7 = ______	2 x 3 = ______
4 x 2 = ______	2 x 10 = _____	2 x 9= ______	2 x 4 = ______
7 x 2 = ______	1 x 2 = ______	2 x 6 = ______	6 x 2 = _____

Level H

H

Write in a word sentence:

10 x 3 ______________________________

Draw a picture: 4 x 3

Draw an array: 2 x 5

Write as an addition problem (use picture above if you need help):

Example 4 x 2 = 2 + 2 + 2 + 2

4 x 3 _____________________________________

New Facts: 3 x 4   4 x 3   3 x 10   10 x 3

H

2 x 7 = ______	4 x 3 = ______	4 x 2 = ______	2 x 10 = _____
3 x 10 = _____	5 x 2 = ______	3 x 2 = ______	7 x 1 = ______
3 x 3 = ______	2 x 8 = ______	10 x 3 = _____	2 x 2 = ______
3 x 4 = ______	2 x 9 = ______	1 x 4 = ______	2 x 4 = ______
6 x 2 = ______	3 x 4 = ______	2 x 3 = ______	7 x 2 = ______

Level I

Write in a word sentence: 4 x 8 ______________________________	Draw a picture: 2 x 5 Draw an array: 3 x 6

I

Write as an addition problem:

Example 6 x 2 = 2 + 2 + 2 + 2 +2 + 2

3 x 6 = ____________________________________

New Facts: 3 x 6   6 x 3   3 x 8   8 x 3

I

3 x 3 = ______	3 x 6 = ______	1 x 4 = ______	2 x 9 = _____
2 x 2 = _____	3 x 6 = ______	3 x 4 = ______	2 x 5 = ______
10 x 2 = _____	6 x 3 = ______	6 x 2 = _____	3 x 2 = ______
3 x 8 = ______	3 x 6 = ______	3 x 8 = ______	8 x 3 = ______
7 x 2 = ______	1 x 9 = ______	4 x 3 = ______	3 x 10 = ______

Level J

J

Write in a word sentence:

10 x 2 ______________________________

Draw an array:

5 x 3

Write as an addition problem:

Example 4 x 3 = 3 + 3 + 3 + 3

3 x 5 = ____________________________________

New Facts: 3 x 5 5 x 3 3 x 9 9 x 3

J

6 x 3 = ______	7 x 2 = ______	1 x 8 = ______	9 x 3 = ______
3 x 10 = _____	2 x 2 = ______	5 x 3 = ______	2 x 4 = ______
2 x 9 = ______	5 x 2 = ______	3 x 4 = ______	3 x 4 = ______
3 x 5 = ______	10 x 2 = _____	3 x 3 = ______	6 x 2 = ______
3 x 2 = ______	9 x 3 =______	3 x 6 = ______	2 x 8 = ______

Level K

K

Write in a word sentence:

2 x 7 ______________________________

Draw a picture:

3 x 7

Write as an addition problem:

Example 3 x 7 = 7 + 7 + 7

4 x 4 = ___________________________________

New Facts: 3 x 7 7 x 3 4 x 4

K

6 x 0 = _____	3 x 7 = ______	9 x 3 =______	3 x 8 = ______
2 x 8 = ______	4 x 4 = ______	7 x 2 = ______	4 x 4 = ______
3 x 3 = ______	3 x 6 = _____	7 x 3 = ______	5 x 3 = ______
2 x 4 = ______	2 x 9 = _____	5 x 2 = ______	4 x 4 = ______
3 x 4 = ______	3 x 10 =_____	10 x 2 = _____	7 x 3 = ______

Level L

L

Write in a word sentence:

9 x 4 ________________________________

Draw an array:

5 x 4

Write as an addition problem:

Example 5 x 2 = 2 + 2 + 2 + 2 + 2

4 x 5 = ________________________________

New Facts: 4 x 5 5 x 4 4 x 9 9 x 4

L

4 x 0 = ______	5 x 4 = ______	2 x 10 = ______	9 x 4 = ______
7 x 3 = ______	5 x 2 = ______	3 x 3 = ______	9 x 3 = ______
3 x 5 = ______	2 x 9 = ______	4 x 4 = ______	6 x 2 = ______
4 x 5 = ______	3 x 1 = ______	3 x 8 = ______	4 x 2 = ______
4 x 9 = ______	3 x 4 = ______	3 x 10 = ______	6 x 3 = ______

Level M

M

Write in a word sentence:

9 x 3 ______________________________

Draw a picture:

6 x 4

Write as an addition problem:

Example 4 x 2 = 2 + 2 + 2 + 2

4 x 5 = ____________________________________

New Facts: 4 x 6 6 x 4 4 x 10 10 x 4

M

4 x 4 = ______	4 x 9 = ______	3 x 4 = ______	4 x 10 = ____
6 x 3 = _____	7 x 0 = ______	5 x 4 = ______	2 x 3 = ______
3 x 5 = _____	7 x 3 = ______	4 x 6 = ______	10 x 4 = _____
9 x 3 =______	3 x 8 = _____	2 x 2 = ______	3 x 10 = _____
8 x 2 = ______	1 x 5 = ______	6 x 4 = ______	2 x 7 = ______

Level N

N

Write in a word sentence:

4 x 7 ____________________

Draw an array:

5 x 2

Division is the opposite of multiplication. Divide into 3 equal groups.

12 ÷ 3 = ______ (how many in each group)

X X X X X X X X X X X X

3 x 4 = 12

Write as an addition problem:

4 x 6 = ________________________________

New Facts: 4 x 7 7 x 4 4 x 8 8 x 4 **N**

3 x 6 = ______	8 x 3 = ______	7 x 1 = ______	7 x 4 = ______
6 x 4 = _____	4 x 7 = ______	4 x 9 = ______	4 x 4 = ______
2 x 10 = ____	6 x 2 = _____	7 x 3 = ______	8 x 4 = ______
3 x 3 = ______	3 x 5 = _____	3 x 4 = ______	4 x 5 = ______
10 x 4 = _____	9 x 3 =______	4 x 8 = _____	0 x 9 = ______

Level O

Write in a word sentence: 9 x 5 ____________________________	Draw a picture: 5 x 3

O

Division is the opposite of multiplication. Divide into 2 equal groups.

10 ÷ 2 = ______ (how many in each group)

X X X X X X X X X X

2 x 5 = 10

Write as an addition problem:

4 x 5 = ____________________________________

New Facts: 5 x 5 5 x 9 9 x 5

O

7 x 3 = ______	3 x 4 = ______	9 x 5 = ______	4 x 5 = _____
4 x 4 = ______	2 x 2 = _____	8 x 4 = _____	3 x 3 =______
5 x 5 = _____	0 x 2 = ______	4 x 9 = ______	6 x 3 = ______
2 x 5 = ______	9 x 1 = ______	10 x 4 = _____	4 x 7 = ______
5 x 9 = ______	5 x 5 = ______	3 x 10 = ____	6 x 4 = ______

Level P

P

Write in a word sentence:

5 x 10 ______________________________

Draw an array:

2 x 5

Division is the opposite of multiplication. Divide into 3 equal groups.

15 ÷ 3 = ______ (how many in each group)

X X X X X X X X X X X X X X X

3 x 5 = 15

Write as an addition problem:

5 x 8 = ____

New Facts: 5 x 6 6 x 5 5 x 8 8 x 5

P

2 x 8 = ______	8 x 5 = ______	9 x 4 = ______	5 x 5 = ______
0 x 8 = ______	3 x 9 = ______	6 x 5 = ______	2 x 7 = ______
8 x 3 = ______	8 x 4 = ______	4 x 7 = ______	5 x 9 = ______
5 x 8 = ______	4 x 10 = ____	6 x 4 = ______	4 x 1 = ______
5 x 4 = ______	5 x 3 = ______	5 x 6 = ______	4 x 4 = ______

Q

Write in a word sentence:

9 x 4 ______________________________

Draw a picture:

3 x 4

Division is the opposite of multiplication. Divide into 4 equal groups.

8 ÷ 4 = ______ (how many in each group)

X X X X X X X X

4 x 2 = 8

Write as an addition problem:

4 x 7 = ______________________________________

New Facts: 5 x 7 7 x 5 5 x 10 10 x 5 **Q**

5 x 0 = ______	5 x 7 = ______	8 x 3 = ______	5 x 5 = ______
3 x 3 = ______	2 x 9 = ______	8 x 5 = ______	9 x 4 = ______
9 x 5 = ______	10 x 5 = ______	3 x 10 = ______	6 x 5 = ______
1 x 6 = ______	7 x 5 = ______	7 x 4 = ______	4 x 4 = ______
5 x 10 = ______	4 x 8 = ______	4 x 6 = ______	2 x 4 = ______

Level R

R

Write in a word sentence:

6 x 9 ______________________________

Draw an array:

2 x 5

Division is the opposite of multiplication. Divide into 2 equal groups.

14 ÷ 2 = ______

X X X X X X X X X X X X X X

7 x 2 = _____

Write as an addition problem:

5 x 3 = ______________________________________

New Facts: 6 x 6   6 x 9   9 x 6

R

3 x 0 = ______	5 x 9 = ______	6 x 9 = ______	5 x 6 = ______
2 x 3 = ______	8 x 1 = ______	4 x 5 = ______	4 x 3 = ______
6 x 6 = ______	4 x 6 = ______	2 x 9 = _____	8 x 5 = _____
9 x 6 =______	9 x 4 = _____	5 x 5 = ______	3 x 9 = ______
6 x 6 = ______	10 x 5 = _____	7 x 5 = ______	8 x 4 = ______

Level S

S

Write in a word sentence:

____

Draw a picture:

4 x 3 5 x 8

Division is the opposite of multiplication. Divide into 4 equal groups.

16 ÷ 4 = ______

X X X X X X X X X X X X X X X X

4 x 4 = _____

Write as an addition problem:

4 x 7 = ________________________________

New Facts: 6 x 7 7 x 6 6 x 10 10 x 6

S

2 x 1 = ______	6 x 6 = ______	6 x 5 = ______	7 x 6 = ______
3 x 6 = ______	6 x 10 = _____	5 x 7 = ______	10 x 0 = _____
10 x 6 = _____	2 x 7 = ______	5 x 5 = ______	4 x 5 = ______
7 x 3 = ______	6 x 7 = ______	4 x 6 = ______	2 x 8 = ______
9 x 5 = ______	8 x 5 =______	4 x 4 = ______	3 x 5 = ______

Level T

T

Write in a word sentence:

7 x 7 ______________________________

Draw an array:

3 x 6

Fact Family – a group of three numbers that make 2 multiplication and 2 division problems.

Facts for Fact Family of 3, 5, 15 (fill in the blanks with missing numbers)

3 x 5 = ______
5 x 3 = ______
15 ÷ 3 = ______
15 ÷ 5 = ______

Write as an addition problem:

3 x 6 = ___________________________________

New Facts: 7 x 7 6 x 8 8 x 6

T

8 x 6 = ______	9 x 6 = ______	9 x 3 = ______	6 x 8 = ______
5 x 8 = ______	1 x 0 = ______	4 x 4 = ______	6 x 5 = ______
7 x 7 = ______	3 x 6 = ______	5 x 7 = ______	5 x 9 = ______
1 x 10 = ____	10 x 6 = _____	2 x 4 = ______	6 x 6 = ______
4 x 7 = ______	7 x 7 = ______	6 x 7 = ______	9 x 4 = ______

Level U

U

Write in a word sentence:

10 x 9 ______________________________

Draw a picture:

3 x 7

Fact Family – a group of three numbers that make 2 multiplication and 2 division problems.

Facts for Fact Family of 4, 6, 24 (fill in the blanks with missing numbers)

4 x 6 = _______
6 x _____ = _______
24 ÷ 4 = _______
24 ÷ 6 = _______

Write as an addition problem:

9 x 6 = ____

New Facts: 7 x 8   8 x 7   9 x 10   10 x 9

U

9 x 5 = _______	4 x 8 = _______	7 x 8 = _______	6 x 7 = _______
5 x 5 = _______	6 x 8 = _______	9 x 10 = ______	7 x 3 = _______
3 x 4 = ______	10 x 6 = ______	5 x 10 = ______	6 x 9 = _______
6 x 0 = _______	8 x 7 = _______	6 x 6 = _______	7 x 7 = _______
5 x 4 = ________	1 x 7 = _______	10 x 9 = ______	2 x 5 = _______

Level V

V

Write in a word sentence:

9 x 6 ___________________________

Draw an array:

4 x 5

Fact Family – a group of three numbers that make 2 multiplication and 2 division problems.

Facts for Fact Family of 3, 6, 18 (fill in the blanks with missing numbers)

3 x _____ = ______

6 x 3 = ______

18 ÷ 3 = ______

18 ÷ 6 = ______

Write as an addition problem:

8 x 8 = __________________________________

New Facts: 7 x 9 9 x 7 8 x 8

V

7 x 9 = ______	6 x 4 = _______	0 x 8 = ______	2 x 9 = ______
6 x 5 = ______	9 x 6 = ______	8 x 7 = ______	8 x 8 = ______
6 x 6 = ______	7 x 7 = ______	10 x 4 = ____	9 x 10 = _____
6 x 3 =______	9 x 7 = ______	8 x 6 = ______	2 x 1 = ______
5 x 9 = ______	5 x 5 = ______	7 x 6 = ______	8 x 8 = ______

Level W

W

Write in a word sentence:

6 x 5 ______________________________

Draw a picture:

2 x 8

Fact Family – a group of three numbers that make 2 multiplication and 2 division problems.

Facts for Fact Family of 4, 5, 20 (fill in the blanks with missing numbers)

4 x ____ = ______

5 x ____ = ______

20 ÷ 4 = ______

20 ÷ 5 = ______

Write as an addition problem:

8 x 9 = ______________________________________

New Facts: 7 x 10 10 x 7 8 x 9 9 x 8 **W**

10 x 6 = ____	7 x 8 = ______	7 x 6 = ______	7 x 7 = ______
8 x 9 = ______	7 x 10 = _____	6 x 9 = _______	0 x 3 = ______
5 x 9 = ______	9 x 10 = ______	9 x 7 = ______	8 x 1 = ______
9 x 8 = ______	8 x 8 = ______	2 x 7 = ______	5 x 3 = ______
7 x 10 = _____	7 x 5 = ______	4 x 7 = ______	4 x 6 = ______

Level X

X

Write in a word sentence:

6 x 3 _______________

Draw an array:

4 x 7

Fact Family – a group of three numbers that make 2 multiplication and 2 division problems.

Facts for Fact Family of 5, 7, 35 (fill in the blanks with missing numbers)

5 x _____ = ______

_____ x _____ = ______

35 ÷ 5 = ______

35 ÷ 7 = ______

Write as an addition problem:

2 x 5 = ________________________________

New Facts: 8 x 10   10 x 8   9 x 9   10 x 10 **X**

2 x 6 = ______	7 x 10 = _____	9 x 9 = ______	8 x 8 = ______
6 x 7 = ______	5 x 4 = ______	10 x 10 = ____	3 x 8 = ______
7 x 7 = ______	8 x 9 = ______	5 x 10 = ______	6 x 1 = _______
10 x 8 = _____	6 x 8 = ______	7 x 10 = ______	9 x 7 = ______
8 x 10 = ______	9 x 8 = ______	5 x 8 = ______	8 x 7 = ______

Level Y

Y

Write in a word sentence:

6 x 7 ______________________________

Draw a picture:

4 x 3

Fact Family – a group of three numbers that make 2 multiplication and 2 division problems.

Facts for Fact Family of 4, 8, 32 (fill in the blanks with missing numbers)

_____ x _____ = ______

_____ x _____ = ______

32 ÷ 4 = ______

32 ÷ 8 = ______

Write as an addition problem:

7 x 4 = ________________________________

New Facts: None

Y

7 x 7 = ______	8 x 9 = _______	3 x 5 = ______	4 x 8 = ______
3 x 7 = ______	2 x 3 = _______	7 x 10 = _____	9 x 9 =______
8 x 9 = ______	1 x 8 = ______	5 x 6 = ______	0 x 9 = ______
2 x 2 = ______	6 x 7 = ______	8 x 8 = ______	4 x 10 = _____
8 x 5 = ______	6 x 9 = ______	10 x 10 = _____	6 x 10 =______

Level Z

Z

Write in a word sentence:

4 x 8 ______________________

Draw an array:

2 x 6

Fact Family – a group of three numbers that make 2 multiplication and 2 division problems.

Facts for Fact Family of 2, 8, 16 (fill in the blanks with missing numbers)

____ x ____ = ______

____ x ____ = ______

16 ÷ 2 = ______

16 ÷ 8 = ______

Write as an addition problem:

8 x 6 = ________________________________

New Facts: None

Z

10 x 10 = _____	8 x 6 = _______	6 x 6 = _______	4 x 4 = ______
3 x 8 = _______	7 x 7 = _______	8 x 8 = _______	9 x 8 = ______
9 x 8 = _______	3 x 6 = _______	2 x 6 = _______	0 x 10 = _____
9 x 9 =_______	5 x 3 = _______	4 x 9 = _______	4 x 6 = ______
5 x 7 = _______	9 x 2 = _______	9 x 7 = _______	8 x7 = ______

Mastery 1

10 x 10 = _____	8 x 6 = _____	6 x 6 = _____	4 x 4 = _____	3 x 8 = _____
7 x 7 = _____	8 x 8 = _____	9 x 8 = _____	0 x 8 = _____	3 x 6 = _____
2 x 6 = _____	0 x 10 = _____	9 x 9 = _____	5 x 3 = _____	4 x 9 = _____
4 x 6 = _____	5 x 7 = _____	9 x 2 = _____	1 x 7 = _____	4 x 0 = _____
3 x 7 = _____	8 x 4 = _____	3 x 1 = _____	0 x 0 = _____	3 x 3 = _____
2 x 3 = _____	3 x 10 = _____	9 x 1 = _____	8 x 9 = _____	1 x 8 = _____
5 x 6 = _____	0 x 9 = _____	2 x 2 = _____	6 x 7 = _____	8 x 8 = _____
4 x 10 = _____	8 x 5 = _____	6 x 9 = _____	10 x 5 = _____	6 x 10 = _____
2 x 6 = _____	3 x 10 = _____	5 x 9 = _____	5 x 5 = _____	4 x 7 = _____
2 x 4 = _____	0 x 2 = _____	3 x 8 = _____	1 x 1 = _____	7 x 0 = _____
5 x 4 = _____	3 x 1 = _____	10 x 8 = _____	0 x 3 = _____	7 x 2 = _____
9 x 7 = _____	5 x 0 = _____	10 x 8 = _____	2 x 7 = _____	10 x 9 = _____
1 x 1 = _____	9 x 3 = _____	6 x 1 = _____	5 x 4 = _____	4 x 3 = _____

Mastery 2

1 x 7 = ______	8 x 8 = ______	9 x 8 = ______	6 x 8 = ______	3 x 6 = ______
2 x 6 = ______	0 x 10 = _____	9 x 9 = ______	5 x 7 = ______	4 x 9 = ______
4 x 4 = ______	5 x 3 = ______	9 x 2 = ______	7 x 7 = ______	4 x 0 = ______
10 x 10 = _____	8 x 1= ______	6 x 6 = ______	4 x 8 = ______	3 x 8 = _____
3 x 7 = ______	6 x 4 = ______	3 x 1 = ______	0 x 0 = ______	3 x 10 = ____
2 x 3 = ______	3 x 3 = ______	9 x 1 = ______	5 x 9 = ______	6 x 8 = _____
5 x 5 = _____	0 x 9 = ______	2 x 2 = ______	6 x 7 = ______	8 x 8 = _____
4 x 10 = _____	4 x 5 = ______	6 x 9 = ______	10 x 5 = _____	6 x 10 = _____
2 x 6 = ______	3 x 10 = _____	8 x 9 = ______	5 x 6 = ______	0 x 7 = _____
2 x 4 = ______	0 x 2 = ______	3 x 8 = ______	3 x 1 = ______	7 x 2 = _____
5 x 8 = ______	1 x 1 = ______	6 x 8 = ______	0 x 3 = ______	7 x 4 = _____
9 x 7 = ______	5 x 0 = ______	10 x 8 = _____	2 x 7 = ______	10 x 9 = ____
1 x 1 = ______	9 x 3 = ______	6 x 1 = ______	5 x 4 = ______	4 x 3 = _____

Mastery 3

8 x 3 = _____	3 x 10 = _____	9 x 1 = _____	6 x 9 = _____	1 x 8 = _____
5 x 6 = _____	0 x 9 = _____	2 x 2 = _____	6 x 2 = _____	8 x 8 = _____
4 x 10 = _____	5 x 5 = _____	8 x 9 = _____	10 x 5 = _____	6 x 10 = _____
7 x 0 = _____	3 x 10 = _____	4 x 4 = _____	8 x 5 = _____	4 x 7 = _____
2 x 4 = _____	0 x 2 = _____	3 x 2 = _____	8 x 8 = _____	7 x 6 = _____
10 x 10 = _____	1 x 7 = _____	6 x 6 = _____	7 x 5 = _____	3 x 8 = _____
7 x 7 = _____	1 x 1 = _____	7 x 5 = _____	0 x 4 = _____	3 x 6 = _____
2 x 6 = _____	0 x 10 = _____	0 x 0 = _____	5 x 3 = _____	4 x 9 = _____
4 x 6 = _____	5 x 9 = _____	9 x 2 = _____	4 x 8 = _____	8 x 0 = _____
3 x 7 = _____	6 x 4 = _____	3 x 0 = _____	9 x 9 = _____	3 x 3 = _____
5 x 4 = _____	3 x 1 = _____	10 x 8 = _____	1 x 3 = _____	7 x 9 = _____
2 x 7 = _____	5 x 0 = _____	10 x 8 = _____	1 x 1 = _____	10 x 9 = _____
7 x 8 = _____	9 x 3 = _____	6 x 1 = _____	5 x 4 = _____	4 x 3 = _____

Mastery 4

4 x 6 = ______	5 x 7 = ______	9 x 2 = ______	1 x 7 = ______	4 x 0 = ______
3 x 7 = ______	8 x 4 = ______	3 x 1 = ______	0 x 0 = ______	3 x 3 = ______
2 x 3 = ______	3 x 10 = ______	9 x 1 = ______	8 x 9 = ______	1 x 8 = ______
5 x 6 = ______	0 x 9 = ______	2 x 2 = ______	6 x 7 = ______	8 x 8 = ______
4 x 10 = ______	8 x 5 = ______	6 x 9 = ______	10 x 5 = ______	6 x 10 = ______
2 x 6 = ______	3 x 10 = ______	5 x 9 = ______	5 x 5 = ______	4 x 7 = ______
10 x 10 = ______	8 x 6 = ______	6 x 6 = ______	4 x 4 = ______	3 x 8 = ______
7 x 7 = ______	8 x 8 = ______	9 x 8 = ______	0 x 8 = ______	3 x 6 = ______
2 x 6 = ______	0 x 10 = ______	9 x 9 = ______	5 x 3 = ______	4 x 9 = ______
2 x 4 = ______	0 x 2 = ______	3 x 8 = ______	1 x 1 = ______	7 x 0 = ______
5 x 4 = ______	3 x 1 = ______	10 x 8 = ______	0 x 3 = ______	7 x 2 = ______
9 x 7 = ______	5 x 0 = ______	10 x 8 = ______	2 x 7 = ______	10 x 9 = ______
1 x 1 = ______	9 x 3 = ______	6 x 1 = ______	5 x 4 = ______	4 x 3 = ______

Mastery 5

2 x 4 = _____	0 x 2 = _____	3 x 8 = _____	1 x 1 = _____	7 x 0 = _____
5 x 4 = _____	3 x 1 = _____	10 x 8 = _____	0 x 3 = _____	7 x 2 = _____
9 x 7 = _____	5 x 0 = _____	10 x 8 = _____	2 x 7 = _____	10 x 9 = _____
1 x 1 = _____	9 x 3 = _____	6 x 1 = _____	5 x 4 = _____	4 x 3 = _____
1 x 10 = _____	8 x 6 = _____	6 x 6 = _____	4 x 4 = _____	3 x 8 = _____
7 x 7 = _____	8 x 8 = _____	9 x 8 = _____	0 x 8 = _____	3 x 6 = _____
2 x 6 = _____	0 x 10 = _____	9 x 9 = _____	5 x 3 = _____	4 x 9 = _____
4 x 6 = _____	5 x 7 = _____	9 x 2 = _____	1 x 7 = _____	4 x 0 = _____
3 x 7 = _____	8 x 4 = _____	3 x 1 = _____	0 x 0 = _____	3 x 3 = _____
2 x 3 = _____	3 x 10 = _____	9 x 1 = _____	8 x 9 = _____	1 x 8 = _____
5 x 6 = _____	0 x 9 = _____	2 x 2 = _____	6 x 7 = _____	8 x 8 = _____
4 x 10 = _____	8 x 5 = _____	6 x 9 = _____	10 x 5 = _____	6 x 10 = _____
2 x 6 = _____	3 x 10 = _____	5 x 9 = _____	5 x 5 = _____	4 x 7 = _____

I am a great believer in luck, and I find the harder I work the more I have of it.

Thomas Jefferson

New Facts Chart

Lesson A: Numbers 1 to 10 times 0

Lesson B: Numbers 1 to 10 times 1

Lesson C: 2x2   2x4   4x2

Lesson D: 2x6   6x2   2x8   8x2

Lesson E:2x3   3x2   2x10   10x2

Lesson F: 2x5   5x2   2x9   9x2

Lesson G:2x7   7x2   3x3

Lesson H:3x4   4x3   3x10   10x3

Lesson I: 3x6   6x3   3x8   8x3

Lesson J: 3x5   5x3   3x9   9x3

Lesson K: 3x7   7x3   4x4

Lesson L: 4x5   5x4   4x9   9x4

Lesson M: 4x6   6x4   4x10   10x4

Lesson N: 4x7   7x4   4x8   8x4

Lesson O: 5x5   5x9   9x5

Lesson P: 5x6   6x5   5x8   8x5

Lesson Q: 5x7   7x5   5x10   10x5

Lesson R: 6x6   6x9   9x6

Lesson S: 6x7   7x6   6x10   10x6

Lesson T: 6x8   8x6   7x7

Lesson U: 7x8   8x7   9x10   10x9

Lesson V: 7x9   9x7   8x8

Lesson W: 7x10   10x7   8x9   9x8

Lesson X: 8x10   10x8   9x9   10x10

Lesson Y: No New Facts

Lesson Z: No New Facts

New Facts Chart

(with answers)

Lesson A: Numbers 1 to 10 times 0	Lesson N: 4x7=28 7x4=28 4x8=32 8x4=32
Lesson B: Numbers 1 to 10 times 1	Lesson O: 5x5=25 5x9=45 9x5=45
Lesson C: 2x2=4 2x4=8 4x2=8	Lesson P: 5x6=30 6x5=30 5x8=40 8x5=40
Lesson D: 2x6=12 6x2=12 2x8=16 8x2=16	Lesson Q: 5x7=35 7x5=35 5x10=50 10x5=50
Lesson E:2x3=6 3x2=6 2x10=20 10x2=20	Lesson R: 6x6=36 6x9=54 9x6=54
Lesson F: 2x5=10 5x2=10 2x9=18 9x2=18	Lesson S: 6x7=42 7x6=42 6x10=60 10x6=60
Lesson G:2x7=14 7x2=14 3x3=9	Lesson T: 6x8=48 8x6=48 7x7=49
Lesson H:3x4=12 4x3=12 3x10=30 10x3=30	Lesson U: 7x8=56 8x7=56 9x10=90 10x9=90
Lesson I: 3x6=18 6x3=18 3x8=24 8x3=24	Lesson V: 7x9=63 9x7=63 8x8=64
Lesson J: 3x5=15 5x3=15 3x9=27 9x3=27	Lesson W: 7x10=70 10x7=70 8x9=72 9x8=72
Lesson K: 3x7=21 7x3=21 4x4=16	Lesson X: 8x10=80 10x8=80 9x9=81 10x10=100
Lesson L: 4x5=20 5x4=20 4x9=36 9x4=36	Lesson Y: No New Facts
Lesson M: 4x6=24 6x4=24 4x10=40 10x4=40	Lesson Z: No New Facts

Let's Continue with Division

Before starting student practice pages work with child on understanding division:

Day 1 (Understanding division is separating the total objects into equal groups)

1. Work with manipulatives to help child understand – use counters, pennies, M & Ms, etc.
2. Start with 15 ÷ 3. Put 15 objects on the table. Make 3 empty groups (make groups by using bowls, saucers, circles drawn on a sheet of paper, etc.). Have child count all objects and write 15 on a piece of paper (because that is the total number of objects). Ask the child how many groups – right 3 "So, we are dividing by 3. We have 15 objects and we are separating or dividing by 3 or into 3 groups. Next have your child write the division sign next to the 15. Then write the 3, followed by the equals sign. Have your child read the problem (15 divided by 3 equals) Prompt child if necessary. Finally, put the 15 objects into the groups. (if this is difficult for your child have him put one in each group. Repeat putting one in each group until the objects are all in the groups. Count the objects. Tell your child the number in each group is the answer. So.15 divided by 3 equals 5. Say, "The math equation to go with this sentence is 15 ÷ 3 = 5." (Be sure your child knows the meaning of the word division and the meaning of the word equation.)
3. If we make five groups instead of 3, how many will be in each group (if necessary, prompt by saying the same number must be in each group). That's right, 3. If we put our 15 objects into 5 groups, each group will have 3 objects. So, 5 x 3 = 15. Draw a picture of both facts (15 objects in 3 groups and 15 objects in 5 groups. Explain to your child that it is the same number of objects, just grouped differently).
4. Play around with some other problems – **3 x 4** or **2 x 6** or **3 x 6** (keep the numbers small, so your child is not struggling with other concepts and can focus on multiplication). When teaching understanding, it is best to use small, easily manipulated facts. This is so your child understands what is happening when he or she multiplies. Understanding is the difference between rote memorization and Common Core goal of fluency.
5. Introduce **fact families.** Work with 3 x 5. Tell child that there are families of facts. A fact family is a group of three numbers that can make 2 multiplication facts and 2 division facts. So, 3, 5, and 15 are a family. Draw your child's attention to the fact that the multiplication problems end with the largest number because multiplication finds the total. The division problems start with the largest number because division is separating the total into equal groups.

   3 x 5 = 15
   5 x 3 = 15
   15 ÷ 3 = 5
   15 ÷ 5 = 3

Note: If you child needs to spend more that one day on any of these lessons, it is better to spend the time now, than rush to get through the material. The goal is a solid foundation. For younger children it might be good to do points 1 and 2 the first day. Point 3 on the 2nd day. Point 4 on the 3rd day. Point 5 on the 4th day.

Day 2 (Multiplication is Repeated Addition)

5. Put the counters in 3 groups of 5. Ask your child if he or she can write an addition problem. If not, show him or her - 5 in this group + 5 in this group + 5 in this group or 5 + 5 + 5 = 15, which gives us the same answer as 3 x 5 = 15.
6. Repeat with 3 x 4. This time your child should be able to write the addition problem of 4 + 4 + 4.
7. Try with 2 x 5 and 2 x 6.
8. Review on Day 3 if your child seems to struggle with the concept.

Day 3 (Zero Property of Multiplication)

5. Work with numbers times 0.
6. 3 x 0 (Explain: You are going to have 3 groups with zero (nothing) in the groups. Ask you child how many objects. Some children struggle with this as they think there has to be some if you are asking. Simply explain, it is zero. We have no objects.
7. Try with 2 x 0.
8. Now tell your child there will be 5 objects in each group, but you don't have any groups. Write the problem for him or her as you say, 'zero groups with 5 objects in each (0 x 5). We have no objects because we have no groups. Sometimes this is easier to understand using concrete objects, such as gum, candy, or cars. For example: 5 Matchbox cars come in a pack. How many Matchbox cars do we have on the table (be sure you have removed all cars from the table). Right, we have zero (zero is the number that represents none) because we do not have any boxes of Matchbox cars on the table. The problem is 0 boxes times 5 cars in a box equals 0. Likewise, the equation for this problem is 0 x 5 = 0.

Day 4 (Identity Property of Multiplication)

3. Teach the identity factor in multiplication. Have your child say, **identity property and identity factor** (make sure your child knows what a factor is). Talk about what the word identity means. Your identity is (mom, dad, teacher, etc). Ask what his or her identity is (prompt with his or her name, if necessary). Then write the letter A on paper. The identity of that letter is a. Its name is A. Do a couple more letters and then switch to numbers. Once your child knows the identity of a number is simply what it is called, you can move to the next step.
4. Tell your child, in multiplication there is a number called the **identity factor** - when you multiply a number by the **identity factor**, you get that same number. For example, 1 x 3 = 3 (use the manipulatives). Do a couple more numbers then ask, what do you think 1 x

53 is? If your child is struggling with this idea, make 1 circle explaining you have 1 group. Then write the number 53 in the circle. Ask how many in the group? Some children will generalize quickly, others need to walk through the concept. Do as many 2-digit numbers as you need in order for your child to understand. Then move to 3-digit numbers. Make a big deal of doing really big numbers, like 3 million, 24 billion, etc. This concept is known as the **Identity Property of Multiplication).**

Day? (Rule of 10)

(There is no day here because you should come back to this once you feel your child has a firm grasp of the concept of multiplication and knows quite a few facts including 4 x 10.)

7. Ask your child what the multiples of ten are (you are looking for 10, 20, 30, 40, 50). You might need to teach the definition of multiple. Use circles with the number 10 in each if you need a concrete example. Then count by 10s, pointing to each circle as you count.
8. Ask what is 1 x 10; 2 x 10; 3 x 10; 4 x 10. Write these problems going down on a sheet of paper, one under the other. Circle all the answers. Ask your child how the answers are alike (they all have zeros). There is a pattern. If your child suggests another pattern besides the fact that they all have zeros, just prompt them to see what other patterns they can see.
9. Look at each answer; and then talk about the problem. What do you think 8 x 10 will be? (Hint for your child: what number was in all the answers – right, zero). Do you think this answer will have a zero? What other number do you think it will have if 1 times 10 is 10 and 2 times 10 is 20. So, 8 times 10 will be? ___
10. What is 9 x 10? Right, 90. If your child needs help with 80 and 90, come back to this lesson next week.
11. What do you think 13 times 10 would be? (Hint: we know it is going to have a 13 and a zero, lets write it down). Good, what number do you have? Right, 130. How about 27 – write it if it is easier.
12. The pattern is to add a zero, we know that because the rule for multiplying by 10 is the answer ends with a zero. This lets us figure out really big numbers when we can't count by 10s.

Day? (Rule for 100)

Teach rule for 100. Follow the Rule for multiplying by ten, changing where necessary to fit the pattern for 100 times a number.

Math is like going to the gym for your brain.

Danica McKellar

Vocabulary

Array: a way of **displaying** objects in equal rows and equal columns

Divide: operation that **separates** number into equal groups

Equation: number sentence using numbers and symbols

Factors: **numbers** that are multiplied to give a product (to give total objects)

Multiplication: operation that gives the total number when you **combine equal groups**

Multiples: The result of multiplying a number by a whole number (Multiples of 2 are 2, 4, 6, ...)

Product: **the answer** to a multiplication problem

Quotient: **the answer** to a division problem

Identity Property of Multiplication: the product of any number and 1 is that number

Commutative Property: numbers can be multiplied in any order and the product will be the same

Zero Property of Multiplication: the product of any number and zero is zero.

It is absolutely critical for competitiveness in the United States for us to raise the bar in education, especially in math, in science, in technology.

Safra Catz

Alternate Strategies

If you need to use alternate strategies, it is because your child is struggling. If that is the case, it is even more important to try to make it fun. People, especially children, put more effort into the things they enjoy. Avoid power struggles at all cost. No one wins.

1. Games are always a favorite. There are many computer games to practice facts. Just google multiplication facts. You can purchase games like Multiplication War from Amazon or a teachers' supply store.
2. Write silly sentences about a hard to remember fact and say them throughout the day. (Example: 8 x 2 = 16) Eight elephants each tried to wear 2 green hats. Can you believe that is 16 green hats. 8 x 2 = 16!)
3. Draw a picture using the equation – put eyes on the 8, 2, and the 16. Repeat the fact. Make animals or cars out of the numbers. Then repeat the fact.
4. Flash cards (Make it fun. Put a hard to remember fact on a card without the answer. Put a couple easy facts on other cards.) Lay out cards one at a time. Have child say the problem and the answer. He gets the ones he knows. You get the ones he needs to practice. See who gets the most. Go through the unknown ones and use strategies to figure the answer.

How to Use this Book in the Classroom

Key Concepts for Teachers:

- Fun, fun, fun: do all you can to keep this fun - rewards, competitions, charts. (Charts and competitions should be respectful of non-ci=competitive students. Charts should be personal, not posted on the wall to avoid embarrassment of students who are struggling.
- A child needs to pass one level before moving to the next level – high five, sticker, smiley face – acknowledge each accomplishment.
- Teach new facts the day before student completes the student practice page, this is the time to figure out answer using understanding of multiplication concepts. This can be homework for the night.
- Student should know facts from memory when they do the student practice page (Lessons A – Z and Mastery 1 – 5)
- If student is on any level for more than 2 or 3 days, practice new facts in isolation (Note: it more effective to practice for 2 minutes 5 times a day than it is to do 1 ten-minute practice session.)
- Students can complete in writing or orally, though this is less likely to be a possibility in the classroom unless you have a co-teacher or a classroom assistant.
- Teach vocabulary – if you need more help, check out *Teaching Vocabulary for Common Core Multiplication and Division*
- One-size does not fit all, use additional strategies if you have students who are still struggling.

The top section of student practice page (Lessons A – Z) is for understanding. The bottom section is memorization and should be timed. Depending on the age and manual dexterity of your child, anywhere from 20 – 30 seconds should be more than enough. The time is not a hard fast rule but intended to keep students focused. Students can also answer orally. Although, one thing to note, research has shown that the physical act of writing triggers the memory area of the brain. Unless a student has a physical handicap, which makes writing difficult, writing will make the memorization process faster.

You will need to set up some type of system to keep track of which students are at any given level. You need to grade the assignments. This is not as overwhelming as it seems given the fact that incomplete sheets do not need to be graded. Have students self-check to be sure their completed facts are correct.

One idea I use is to have students complete in pen, then self-check. You would only need to verify the students who passed the level. Once you get passed the early levels, there are not as many students completing the page every day.

Key Ideas (teach these concepts) **for Students in the Classroom:**

- Multiplication problems can be reversed without changing the answer – 3 x 5 has the same answer as 5 x 3
- 3 x 5 is read as 3 times 5
- 3 x 5 means 3 groups of 5 objects
- Multiplication is repeated addition so 3 x 5 equals 5 objects + 5 objects + 5 objects

  3 groups with 5 objects in each group equals 5 + 5 + 5.

Made in the USA
Monee, IL
08 November 2026

68002501R00033